21st Century Skills Library

LIFE SKILLS BIOGRAPHIES

Larry Page and Sergey Brin

James M. Flammang

Cherry Lake Publishing
Ann Arbor, Michigan

Published in the United States of America by Cherry Lake Publishing
Ann Arbor, MI
www.cherrylakepublishing.com

Content Adviser: Josh McHugh, Contributing Editor, *Wired*, San Francisco, California

Photo Credits: Cover and page 1, © Boris Roessler/dpa/Corbis; page 7, © Peter M. Fisher/Corbis; pages 17, 20, 23, 24, and 43, © Kim Kulish/Corbis; page 18, © Michael Lewis/Corbis; page 26, © Shaul Schwartz/Corbis; pages 31, © Chris Farina/Corbis; page 33, © Mike Segar/Reuters/Corbis; page 35, © Shaffer Tim/Corbis Sygma; page 36, © James Leynse/Corbis; pages 39 and 40, Catherine Karnow/Corbis; page 42, © Christopher Farina/Corbis

Copyright ©2008 by Cherry Lake Publishing
All rights reserved. No part of this book may be reproduced or utilized in any form or by any means without written permission from the publisher.

Library of Congress Cataloging-in-Publication Data

Flammang, James M.
  Larry Page and Sergey Brin / by James M. Flammang..
    p. cm. — (Life skills biographies)
  Includes bibliographical references and index.
  ISBN-13: 978-1-60279-070-4 (hc : alk. paper)
  ISBN-10: 1-60279-070-1 (hc : alk. paper) 1. Brin, Sergey, 1973—Juvenile literature. 2. Page, Larry, 1973—Juvenile literature. 3. Computer programmers—United States—Biography—Juvenile literature. 4. Internet programming—United States—Biography—Juvenile literature. 5. Google—Juvenile literature. I. Title. II. Series.

QA76.2.A2F56 2007
005.1092'2—dc22
  [B]   2007005665

―――――――――――――――――

*Cherry Lake Publishing would like to acknowledge the work of The Partnership for 21st Century Skills.
Please visit www.21stcenturyskills.org for more information.*

Contents

Introduction   4

CHAPTER ONE
An Early Start   5

CHAPTER TWO
Choosing Careers in Information Technology   10

CHAPTER THREE
Google Inc. Is Born and Grows Fast   17

CHAPTER FOUR
Giving Back to the Community   29

CHAPTER FIVE
Google's Legacy and Future Plans   36

Timeline   44

Glossary   46

For More Information   47

Index   48

About the Author   48

Introduction

More than 100 million search requests come to Google each day from more than 100 countries. It is the most popular search engine in the United States. And its founders, Larry Page and Sergey Brin, are billionaires. Their creative approach to organizing information has changed people's expectations of the Web. Their unconventional style of running a company—with the primary goal of solving problems and changing the world for the better, as opposed to making money—has changed how people think about doing business. And they are only in their mid-30s.

They were both working with computers by their ninth birthdays. These **prodigies** have always preferred challenges and trying what hasn't been done. Friends and relatives knew they'd be successful. But the scale of their success is extraordinary. And what they're doing to share their success is equally exceptional.

CHAPTER ONE

An Early Start

Even though they were born on opposite sides of the world, Larry Page and Sergey Brin had a lot in common. Both boys were prodigies. Smarter than most of their schoolmates, both began to achieve at an early age.

Larry and Sergey also started working with computers at a time when few homes and schools had them. Each boy grew up in an academic household with parents who worked at a university or research center. Friends and relatives predicted successful and rewarding careers for both boys, even when they were just in grade school.

Computers can be found in many homes and schools today, but very few homes and schools had computers when Larry Page and Sergey Brin were boys.

Carl Page attended the University of Michigan in Ann Arbor.

Born in March 1973, Lawrence (Larry) Edward Page grew up in East Lansing, Michigan. His father, Carl Victor Page, was a professor of computer science at Michigan State University. In 1960, Carl Page had been one of the first students to earn a degree in computer science from the University of Michigan in Ann Arbor. He was also among the first to teach the subject and is considered a pioneer in such computer-related areas as artificial intelligence, which deals with the simulation of intelligent behavior in computers.

Larry's mother, Gloria Page, also worked professionally with computers. After earning a master's degree in computer science, she worked with many companies to help them improve their computer systems and taught computer programming.

Sergey Mikhailovich Brin was born in August 1973 in Moscow. That city was then the capital of the Soviet Union and is now the capital of Russia. In 1979, his family **emigrated** to the United States to escape **anti-Semitism** in the Soviet Union. Sergey spent his early school years in California and Maryland.

Sergey's father, Michael Brin, was trained as a mathematician and an economist. After moving his family to Maryland, he became a professor at the University of Maryland, where he was highly regarded. Sergey's mother, Eugenia Brin, was a mathematician and engineer. She worked at the NASA Space Flight Center in Maryland

Both boys quickly developed a powerful interest in mathematics and computers. Sergey received his first computer, a Commodore 64, at

Both Sergey's father and mother were trained to solve complicated mathematics equations, such as the one partially shown here.

Learning & Innovation Skills

During their early years, Larry and Sergey attended Montessori schools. Maria Montessori, the first female physician in Italy, developed a method of childhood education in the early 1900s that eventually was named after her. Individual initiative and self-direction are key to her method, which is based on scientific observation of the ways in which each child learns. Montessori believed that children can almost teach themselves. They learn at their own pace in supportive environments. "Teach by teaching, not by correcting" is one of Montessori's guiding principles. How do you think attending Montessori schools prepared Larry and Sergey for their future success?

age nine. It was a birthday gift from his father. Before long, Sergey was writing code to create computer programs.

Larry fell in love with computers when the Page family purchased a home computer in 1978. Unlike computers today, theirs was big and expensive. Larry later said, "I think I was the first kid in my elementary school to turn in a word-processed document."

At Macdonald Middle School in East Lansing, Larry had a special teacher who worked with him and other advanced students. Larry was interested in many subjects and had a highly inquisitive mind.

So did Sergey. Both boys liked to think and study, and figure out how things worked—and not just computers. They were intrigued by everything around them: what politicians did, how government operated, how the nation's economy functioned. And from an early age, they enjoyed debating about current events with their families rather than watching television or playing games.

At age 12, Larry read a biography of Nikola Tesla (1856–1943), who became an inspiration to him. Tesla was a highly regarded scientist and productive inventor. Though he developed a string of notable innovations in electricity and radio, he earned little fame or money during his life. Larry wanted to be an inventor much like Tesla, but he didn't want to end up poor and ignored. The troubled inventor's life taught Larry valuable lessons. He learned that it's essential to consider the commercial applications of anything you invent or develop. If he ever invented something great, he knew to make sure it got into the right hands and that he would get paid for it. "To just invent something and have a great idea is a lot of work," he said later, "but it is not enough. You have to get it out in the world."

Friends and neighbors who knew Larry as a boy recall that he was highly intelligent but quiet and contemplative. He appeared to be thinking all the time.

Sergey is remembered as more of an **extrovert**, but no less brilliant. Even today, he is considered the more outgoing of the two Google founders.

21st Century Content

Born in the Balkan region of Europe in 1856, Nikola Tesla studied mechanical and electrical engineering. While in Europe, he invented an induction motor and worked for electric power companies. When he was 28, he came to the United States. He worked for famed inventor Thomas Edison, but they parted ways after Edison refused to pay him an expected fee. During his long career, Tesla developed solar cells, X-rays, a radio transmitter, and much more. He was granted more than 100 patents in the United States, which gave him exclusive rights to make, use, or sell his inventions. His work contributed to the adoption of alternating current (AC) rather than direct current (DC) as a method of transmitting electricity over long distances.

Overshadowed by Edison, Tesla achieved neither fame nor fortune. Despite Tesla's contributions to wireless communication, Guglielmo Marconi came to be called the Father of Radio and won a Nobel Prize. When Tesla died in 1943, at age 86, he was poor.

CHAPTER TWO

CHOOSING CAREERS IN INFORMATION TECHNOLOGY

Sergey Brin enrolled at the University of Maryland at the age of 16.

Sergey Brin attended Eleanor Roosevelt High School, a public school located near Washington, D.C. While still enrolled there, at age 16, he entered the University of Maryland. At age 19, in 1993, he earned an undergraduate degree, achieving honors in math and computer science. While at the university, he also took many graduate-level courses.

During high school, summer jobs for Larry Page involved computer science and math—his two favorite fields. After graduating from East

Lansing High School, he enrolled at the University of Michigan in Ann Arbor. He earned a bachelor's degree in engineering with an emphasis in computer science, graduating with honors.

Interested in becoming either a professor or an **entrepreneur**, Page decided graduate school would be his next step. He felt the best place to go was Stanford University. The school was located in the region of California that was becoming the heart of the computer world, called Silicon Valley. Stanford had a powerful reputation in the computer field. Countless computer people, including the founders of Yahoo!, had gone to graduate school there.

In March 1995, Page visited Stanford and took a tour for prospective computer science graduate students. Brin was already a student in Stanford's graduate program. He had arrived there in 1993 with a fellowship from the National Science Foundation. Then in his second year at Stanford, Brin was serving as Page's tour guide.

During that first meeting, it was obvious to others on the tour that Brin and Page were brilliant. But they didn't get along that well and argued a lot. Each had strong opinions about all sorts of topics, not just computers or technology. Other subjects, such as urban planning, also provoked lively disagreements.

Their personalities differed a lot, too. Brin was a gregarious young man and liked to meet new people. Page was shy and reserved. At first, he considered the slightly younger Brin to be somewhat annoying.

Because Brin had taken so many graduate-level courses before arriving at Stanford, he didn't find his classes very stimulating. So he spent a lot of

If Page and Brin had let the fact that they didn't like each other at first prevent a relationship, Google probably would not have been invented. But they didn't allow their differences to keep them from working together. People with different personalities are often complements to each other. One person's weakness might be another's strength. It turned out they work well together.

Larry Page and Sergey Brin attended graduate school at Stanford University during the 1990s.

time swimming, diving, and sailing. He even learned how to "fly" on a trapeze, like a circus performer.

Page was eager to begin his graduate work. "I was really excited to get into Stanford," he later told a biographer. "There wasn't any better place.... I always wanted to go to Silicon Valley."

As each realized how capable the other one was and how much they had in common, Brin and

Page grew closer. They began to work together and soon became almost inseparable. Graduate students are assigned office space, and Page's office became their favorite meeting spot.

Page was still wavering between an academic career and one in business. Before long, his interest in becoming a professor began to fade. "I realized I wanted to invent things," he told a biographer in 2005. "But I also wanted to change the world. I wanted to get [my discoveries] out there, get them into people's hands so they can use them, because that's what really matters." Page liked to think on a big scale. As the president of the engineering honor society in college, he proposed a campus monorail system—which was never built. And though he considered several possible topics for his doctoral work, he decided to concentrate on the World Wide Web. He wanted to study it from a mathematical perspective.

Learning & Innovation Skills

In the early days of the World Wide Web, when few sites existed, indexes weren't so essential. As the number of Web sites online increased, Internet users needed a way to search them for the information they wanted.

Search engines use a "crawler" or "spider" program to scan Web sites. Some people call this type of program a robot. At Google, they call it a Googlebot. It searches Web sites and then sends details about each site back to an indexing system.

Search engines are somewhat like catalogs at a library. When you want to find information at the library, you search for subjects in the catalog. The Dewey Decimal System, which librarians use to give each book a unique number, lets you find each one easily on the shelf.

On the Web, the search engine lists Web sites in much the same way. Sites are indexed almost like books at a library. What do you think the World Wide Web would look like without search engines organizing the information?

Page and Brin were fascinated by the World Wide Web and the wealth of information it provided to computer users.

Brin said later that he "tried so many different things in grad school. The more you stumble around," he explained, "the more likely you are to stumble across something valuable."

Page and Brin were especially intrigued by how search engines compiled and organized information from the World Wide Web, which was attracting new users every minute. Several search engines were operating by the mid-1990s, providing information to millions of computer users. But the quality of information that they delivered was not great. The search engines were unable to evaluate the usefulness of the Web sites they found. All they could do was generate long lists of sites that might be relevant.

At Stanford, Page and Brin wanted to create a way for search engines to evaluate and rank the sites they discovered, and to do so quickly. Classmates and colleagues at Stanford didn't think this was a promising area of work. Most of them devoted their efforts to other computer-related concepts and technologies, including improving Internet **browsers**.

Beginning in January 1996, the two collaborated on a project that evolved into a primitive search system called BackRub. It could analyze the "back links" that point to a given Web site. If a site was linked to many other sites, especially popular and important ones, Page and Brin reasoned that it probably contained more valuable information. This was an imaginative way to evaluate the importance of each site relative to others.

Brin and Page named the basic principle of the system PageRank. It is the most vital element in the set of instructions, or **algorithm**, that the two men created for their search engine. Largely Page's creation at first, it served as the basis for both men's doctoral projects.

To further develop their creation, Brin and Page needed a lot of computers. They had determined that searching and indexing the Web would work best using a network of small computers. But, like most graduate students, they were short on cash. And they had already used up most of their money buying about 120 hard drives, which are devices for

Life & Career Skills

Throughout the process of developing Google, Page and Brin demonstrated imagination and persistence. They decided to use many small computers networked together. At the time huge mainframes, which hold vast amounts of information in one place, were more commonly used in search engines. Page and Brin also determined that the popularity of a site is one of the most significant factors in assessing its value. Equally important, they thought hard about what people really wanted and needed from their searches.

Page and Brin envisioned what they wanted their search engine to do and kept working on it until it matched their vision. Perhaps most important, they had to be confident that they would eventually succeed.

storing data. They built their own computer housings for the hard drives in Page's dormitory room. They would even visit the loading docks for the computer science department. If new computers arrived, they'd ask to borrow a few for a while to test their innovative search engine.

In 1998, they wrote a paper called "The Anatomy of a Large-scale Hypertextual Web Search Engine," which outlined their system. In it, they introduced the name Google and its distinctive features, including PageRank. Stanford's own Web site served as the host for the early version of Google. The two young men spent the first half of that year perfecting the technology they'd created.

At this point, they didn't plan to start a company. Soon, however, they began to realize the potential of their work.

CHAPTER THREE

GOOGLE INC. IS BORN AND GROWS FAST

Larry Page (left) and Sergey Brin (right) in the server room at Google Inc. When the two men started the company, they had no idea it would grow so quickly.

When they first developed what would become Google, neither Brin nor Page was thinking seriously about forming a company. As Stanford students and teachers began to use and praise the search engine in 1997, Brin and Page began to suspect some commercial possibilities. When

venture capitalists and product developers in the Silicon Valley area started to talk about it, the two knew they had created something with prospects in the marketplace.

Rather than start a business themselves, Brin and Page tried to sell the Google concept to companies that were already operating search engines and other computer-related organizations. Their attempts to sell the idea, or license its use, proved to be unsuccessful. No one in the business, including the people at Yahoo!, was interested in the idea because they could not see a way to make money with it.

At this time, the most popular search engines on the World Wide Web, such as Alta Vista, Excite, and Yahoo!, were receiving tens of millions of queries per day. Yahoo! and other early search engines used knowledgeable people to catalog the Web. But the Web grew so fast that these experts could not keep pace. Brin and Page believed their system could.

Why did they call it Google? Years earlier, a mathematician named Milton Sirotta came up with the word *googol* to identify a huge number: 1 followed by 100 zeroes. To demonstrate that their search engine could find and index vast numbers of Web sites, Brin and Page decided to use that word. But they spelled it slightly different. Besides, *Google* sounded like fun.

In 1998, as they thought seriously about establishing Google as a business on their own, both put their studies on hold. They had each already earned a master's degree at Stanford, but the **dissertations** they would have to research and write in order to obtain doctorate degrees were set aside. Earning these highest degrees would be essential if they were to become professors. Instead, both took leaves of absence from Stanford's doctoral program. The two young men were ready to concentrate on the commercial prospects for Google.

In the early months of setting up the Google organization, their dormitory rooms at Stanford served as company headquarters. Page's room became the data center, while Brin's residence turned into the company's business office.

Page and Brin knew they'd have to come up with cash to get started. Credit cards were the temporary solution to their lack of funds. Before long, both men had pushed their personal credit cards to their maximums. Friends and parents also offered their credit cards, but that money wouldn't last long. They needed funds from outside investors.

Encouraged by the success of Silicon Valley entrepreneurs such as David Filo and Jerry Yang, who had started Yahoo! in 1995, Brin and Page began letting potential investors know about Google. First to respond was Andy Bechtolsheim, the cofounder of a large computer company called Sun Microsystems, who wrote Google a check for $100,000.

Brin and Page had to put Bechtolsheim's check in a drawer for a couple of weeks though. They didn't have a company bank account, so there was no way to cash the check right away.

Within a few weeks, they raised nearly $1 million. The additional investment money came from family members and friends, as well as people in the computer industry whom they'd met in the Silicon Valley area.

On September 7, 1998, the new company was incorporated as Google Inc., with a staff of three. Page took the title of chief executive officer

Andy Bechtolsheim, cofounder of Sun Microsystems, was one of the first people to invest in Google.

(CEO). Brin was named Google's president. And Craig Silverstein, a fellow Stanford student, became director of technology.

Like many newcomers to the computer field, such as Steve Jobs and Steve Wozniak (who had formed Apple Computer many years earlier), Brin and Page started small. They set up shop in the garage and a few rooms of a friend's house in Menlo Park, California. As 1998 drew to a close, Google was getting 10,000 search queries per day.

Google became a nearly instant success. During its first five years, it grew faster than any other company in the United States.

Amazingly, the company's rapid growth happened without any advertising. Google relied upon word of mouth. One user would simply tell another how Google worked and how helpful the results were. Each new, devoted user might tell several others about it, until thousands knew about Google.

The number of users grew even faster in 1999, when AOL/Netscape decided to use Google as the search engine offered on its own Web site. A year later, Yahoo! partnered with Google, too.

By the spring of 1999, Google had eight employees. Their search engine was getting half a million queries per day. In June of that year, the two founders raised more than $25 million from venture capital groups to expand operations.

After spending several months in a small office space in downtown Palo Alto, they opened an impressive two-story headquarters in nearby Mountain View, California. Named the Googleplex, the building had plenty of space for expansion. By the time it opened, 39 employees

Page and Brin followed a few important principles on their path to success. One principle is to always strive to do better. Though their search engine is extraordinarily popular, both men have remained focused on continually refining it. Another is that whatever you choose to do, do it as well as you possibly can. It's far better to do one thing really well than to do many things poorly. When the company has expanded, the new endeavors have been related to Google and operated in much the same way.

were on the staff and about three million searches were being conducted every day.

Publicity came as fast as the company's growth. Profiles of Brin and Page, who were only in their mid-20s, appeared in *Time* and *Business Week* magazines and in newspapers like *USA Today*.

During its first few years, Google grew at a fantastic rate. By the middle of 2000, 18 million queries per day were coming in. The Google index contained one billion Web pages.

People liked Google not just for its results, but because it worked so fast. Once a user selected a Web site from the listings, the page was quickly loaded onto the user's computer. Rather than trying to keep each user on the site for a long time, Brin and Page wanted their customers to leave Google as soon as possible. That would suggest users had found what they wanted.

Despite it success, however, the company was not making money. New companies often operate for several years before they begin to earn a profit. Unless Google became profitable soon, the company's working capital, obtained from early investors, would be gone.

But how could they earn money? Users sitting at their computers pay nothing for the Google search services. In that way, Google is like a library catalog or a telephone book. This kind of information is free. Brin and Page wanted to keep Google free of charge because they realized that computer users would never want to pay for such a service, no matter how helpful it was.

Instead, Google turned to advertisers, but not in the usual way. At first, Page and Brin wanted no advertisements on the Web site at all, because

The familiar Google logo decorates the signs at the company's headquarters in Mountain View, California.

they considered ads to be annoying. Eventually, they decided to accept them but only on their own terms. They didn't want any ads to appear on Google's home page, which they preferred to keep as simple as possible. And they didn't want pop-up ads on the search pages.

Page (left) and Brin (right) came up with an innovative way for their company to make money.

As an alternative, they offer only simple targeted advertisements, alongside or above the search listings. Ads have only a few words and no graphics of any kind. And they have to be related in some way to whatever the user is trying to find. From the beginning, all Google ads have been labeled as such, so everyone knows which words on a page are paid-for advertising.

Brin and Page knew that advertisers wanted access to all those millions of Google users. After all, every one of those users was looking for something specific—certain products or information. Google knows roughly what you're looking for with each search, so it can put up only ads that are related to your specific interest at that moment.

Because advertisers were so eager to get space on Google, Brin and Page reasoned that they would be willing to accept the requirements for the ads without complaint.

Starting in 2001, advertisers could bid on certain keywords. For example, a company that sold car tires might bid for ad space on pages that come up when someone searches for information on tires. The AdWord program, as this system was called, paid off quickly. Google earned its first profit at the end of 2001.

Google's rapid growth prompted some investors to question whether Brin and Page were the right people to head the company because they had started out with no real business management experience. These concerns were addressed in August 2001, when Eric Schmidt joined Google as CEO. Unlike Brin and Page, Schmidt was a thoroughly experienced executive who had been the head of Novell, a major computer firm. Brin was now Google's president of technology. Page's new title was president of products.

In August 2004, Google Inc. began offering its stock to all interested investors. The initial public offering (IPO) of stock turned into one of the

A stock market video monitor announces the initial public offering of Google stock.

strongest ever. The original asking price for each share of stock was $85. Prices per share continued to grow. By November, stock was selling for more than $200 a share. In mid-2005, the price topped $300.

A second offering of stock took place in August 2005. Late that same year, Google's stock price rose above $400. By then, the company had almost 3,500 employees. More than 200 million search queries came to Google every day.

As part of the IPO, more than 1,000 Google employees

who had been given stock options suddenly became millionaires. Many workers who'd joined Google had accepted relatively low salaries because of the stock options that came with the job. They hoped that one day, as Google became successful, they could exercise those options and their investments would shoot up in value. As things turned out, they were correct.

When a company goes public, its stockholders often largely determine its future. Representatives of the stockholders, such as the corporate officers on the board of directors, normally make decisions based mainly on money. But Brin and Page set up the public offering and structure of the company so that they would still make all the vital decisions themselves, though they consult with others. They wanted to ensure that the focus of the company didn't become solely making money. Brin and Page wanted to continue to emphasize the importance of the principles that have governed the company's operation from the beginning. One is their concern for

Learning & Innovation Skills

In the late 20th century, Silicon Valley in California became the center of the emerging computer industry. Nicknamed Silicon Valley by a journalist in 1972, the area covers most of Santa Clara County, between San Francisco Bay and the city of San Jose.

Silicon is an element in semiconductors, which are the "brains" of computers. Stanford University, which has played a major role in the study of computers, is in the city of Palo Alto. Hewlett-Packard, one of the early makers of personal computers, was among the first companies to settle in the region.

How do you think this area of California became the center of the computer industry? How has the area where you live responded to the growth of the computer industry? Is there a company that makes semiconductors or other computer parts in or near your town?

employees and fostering a team environment. Another is considering the user, or customer, first. They want to give each user the best possible experience, in the shortest possible time.

In September 2005, a survey by *Vanity Fair* magazine named Brin and Page the "most powerful leaders of the information age." They reached the top ranks of *Forbes* magazine's list of the world's richest people.

In less than 10 years, Page and Brin had become billionaires, flying in their own Boeing 767 jet. Unlike many people who become famous and seek publicity, they have remained private men.

By the end of 2006, almost 10,000 people worldwide were working for Google, and the company had bought YouTube, the popular video sharing service.

Page has often spoken of having a "healthy disregard for the impossible." As he told a biographer, "You should try to do things that most people would not."

Because many computer-related companies collapsed during the 1990s and into the 21st century, the huge success of Google might have seemed impossible. Its achievement is a prime example of what can be accomplished with what the Biography Channel called "one extraordinary idea."

CHAPTER FOUR

Giving Back to the Community

The public learned of Page and Brin's desire to share their wealth from letters written to Google shareholders.

By 2007, Brin and Page were each worth more than $14 billion. Like many wealthy people, the two men wanted to find ways to repay their community and their country, if not the whole world. Brin even told a biographer that Google "was the smallest of accomplishments that we hope to make over the next 20 years."

When stock in Google was first offered to the public in 2004, Page and Brin sent a letter to their shareholders, stating that they intended to "use

1 percent of the company's **equity** and profits for **philanthropy**." In fact, they expressed their hope that this would eventually "**eclipse** Google itself in terms of overall world impact."

On October 11, 2005, Sheryl Sandberg, Google's vice president of global online sales and operations, posted a plan for this charitable branch on the company blog. Page and Brin called it Google.org. Right away, they announced that they expected to put $1.2 billion into this venture over 20 years, but they believe that the charitable investment could be much greater by the time that period ends because it is based on stock and is therefore dependent on the stock price.

Unlike most charitable organizations, which are nonprofit, this one is a profit-making enterprise. The company believes this method gives the organization "greater flexibility," because it is not subject to tax-related restrictions by the government. This allows Google.org to provide funds for start-up companies around the world. It could also establish partnerships with venture capitalists. It could even hire lobbyists to promote projects to the U.S. Congress. Any project that makes money will pay taxes on its profits.

In February 2006, Google hired Larry Brilliant as the executive director of Google.org. Brilliant is a physician and public health expert. At one time, he was also an entrepreneur in Silicon Valley. Brilliant explained to an interviewer that Google.org "can start companies, build industries, pay consultants, lobby, give money to individuals, and make a profit."

Though the organization can make money, it's "not doing it for the profit," Brilliant said in an interview with the *New York Times*. "The

Bill Gates and his wife, Melinda, have donated billions of dollars from their personal fortune to help people around the world.

emphasis is on social returns, not economic returns." According to the organization's Web site, Google.org addresses challenges relating to poverty, disease, and climate change. As an epidemiologist who has

Bill Gates, the wealthy founder of Microsoft, takes a more traditional approach to philanthropy. Gates and his wife, Melinda, contribute large sums to charitable groups that fight poverty, hunger, and disease around the world. The nonprofit Bill and Melinda Gates Foundation gives away money from Bill and Melinda Gates's personal fortune. In 2004 alone, the Gates Foundation gave away more than $1 billion.

If you had $1 billion, how would you practice philanthropy?

Learning & Innovation Skills

Innovation is the introduction of a new idea, method, or device. Innovation has been a major part of Google's success. Google has teams of workers who compete to come up with innovative ideas. Page rewards the team with the most innovative idea by giving the team members stock in the company. This means the employees own small pieces of Google. Through this friendly competition, Page has given away several million dollars' worth of Google stock.

Why to you think innovation is so important to a company like Google?

worked on epidemics of serious disease, Brilliant is especially eager to find a way to detect outbreaks before they become life-threatening.

One project that the organization supports is trying to develop an ultra-fuel-efficient hybrid car, which can run on electricity, ethanol, and gasoline. If successful, it could be plugged into an electrical outlet for recharging and go 100 miles on a gallon of fuel. Quite a few hybrid automobiles are already for sale, but they aren't nearly as fuel-efficient as this one might be.

Google.org includes the nonprofit Google Foundation, which was started with a $90 million **endowment**. The Google Foundation's initial commitments included the Acumen Fund, which helps entrepreneurs build businesses that provide goods and services to the poor in developing countries. TechnoServe helps new entrepreneurs get their businesses rolling. The foundation is also supporting research in Kenya, Africa, to improve water quality and a program to improve literacy in India.

Even before Google.org was established, the Google company had begun work on many projects that are meant to benefit society. One project,

A worker at the New York Public Library scans a book page. The library is one of the libraries working with Google to make millions of books available online.

initially called Google Print and now named Google Book Search, involves scanning and digitizing millions of books from libraries. Rather than going to the library to try and find the book you need, you could find it online and search its contents for exactly what you need. Books and other items under **copyright** are only partially shown, but some authors and publishers are not happy with the idea. They claim that they'll lose sales of their books if people can access them online.

Harvard University's library is one of the libraries participating in the Google Book Search project.

Late in 2004, Google Print announced agreements with the libraries of Harvard University, Stanford University, the University of Michigan, and Oxford University, as well as the New York Public Library, to digitize some of their collections.

Brin and Page's goal is definitely ambitious. Basically, they would like to organize all of the world's information, not just Web sites. And they want to make all of it accessible to everyone.

Google also has worked with biologist Craig Venter on genetic and biological research, exploring why certain people become ill or develop allergies. The research focuses on how a person's genes affect the likelihood of contracting certain diseases. The speed and style of Google's search methods could be of great benefit when searching a person's genetic makeup, because there are millions of genes that can be analyzed. The process has even been called "Googling Your Genes."

Google's search engine itself is used for charitable purposes. The Google Grants program gives free advertising to certain nonprofit organizations that are dedicated to community service.

Giving money is only one way to improve people's lives. Google's two founders also give back to the community using a different kind of currency. They have given the world a different way of doing business.

Dr. Craig Venter has worked with Google on the process that has been referred to as "Googling Your Genes."

CHAPTER FIVE

Google's Legacy and Future Plans

What began as the name of a company has become a universally recognized word.

All over the world, computer users know what "to Google" means. Millions of them do so often, whether for schoolwork, business questions, research, or fun. The word *google* is even listed in many dictionaries.

PC Magazine has called Google "an almost frighteningly accurate search engine." Their testing "found that the quality of the results matches or exceeds that of every other site tested."

Perhaps the greatest impact of Google isn't its search engine at all. Instead, Brin and Page's greatest contribution to society might be the way their company operates. In most respects, it differs sharply from a typical large corporation or businesses of any kind. "Don't be evil" is Google's simple but crucial motto, which came from a company engineer. "Learning, knowledge, teaching" were primary Google objectives, according to Terry Winograd, who was Page's mentor at Stanford.

But their ultimate goal was made clear when the company went public in 2004. A letter to prospective investors explained that Google was "not a conventional company." Rather than just a way to make money, it was meant to be "an institution that makes the world a better place." For most companies, money and profits are what matter. Emphasis on earnings has resulted in many financial scandals in the corporate world. Google's founders believe companies can make money—a lot of money—without doing anything unethical.

As Google's Web site explains, few company executives can "resist the temptation to make small sacrifices to increase shareholder value." But the focus at Google is its customers—the computer users of the world—ahead of profits. The company "isn't run for the long-term value of our shareholders," chairman Eric Schmidt told *Time* magazine, "but for the long-term value of our end users."

If a more traditional company made such a statement, it might be criticized as an exaggeration. But nearly everything Google does seems to reinforce the principle that "you can make money without doing evil." As the corporate Web site insists, "Focus on the user and all else will follow."

Treating employees well is at least as important as treating customers properly. Here, too, Google excels. In 2007, *Fortune* magazine named Google the Best Company to Work For in America. Many people describe the Googleplex as more like a university campus than a corporate headquarters. Employees are known as Googlers. When they're seen and interviewed on TV news broadcasts, most of them seem unusually enthusiastic. Although they put in long hours every day, they seem to enjoy every minute. Unlike employees at many companies, they don't appear eager to go home at night. Google employees tend to be young and single, with few family responsibilities. And most of them seem to be doing work they love for a company that encourages innovative thinking.

From the start, Google promoted what it calls a "unique company culture" in an "informal atmosphere." In the early days, rubber exercise balls served as movable office chairs, and desks were wooden doors on sawhorses. Inside the Googleplex today, there

Learning & Innovation Skills

Google has its critics. Advocates of privacy worry because a record of every search, which includes information about the user's computer and browser, as well as the date and time, goes into the huge Google database. They wonder if all that information about people's searches could get into the hands of criminals or others who might misuse it.

Validity is another issue. There is no way to know if information in Google's search listings is true or accurate. Often, too, it's difficult to know whether the information is recent or old.

What do you think Google should do to address these concerns? How could the company ensure users that information gathered in the database won't be misused? What could Google do to make only the most recent search results available to users?

are no cubicles and no divider walls. All employees are out in the open. Many "toys" dot the premises for workers to play with, so they won't mind the long hours and modest salaries. A 24-hour cafeteria provides three healthy meals per day at no cost.

Brin (top) and Page (seated on the floor) enjoy the playground-like environment of the Googleplex.

Googlers take time during their workdays to pursue activities that help them relax.

Free medical and dental care are also available at the Googleplex. So are washing machines. If a worker needs a haircut, a barber or stylist is right there.

During their workday, employees might spend some time playing foosball or working on puzzles. Vibrating massage chairs might ease the aches and pains of long hours at a desk. Google provides a lengthy list of activities and devices to make those long days fun as well as highly productive.

Supervision differs from most companies because there is no conventional **hierarchy**, and employees normally work in teams of two or three. Employees are also encouraged to spend one day a week working on any ideas or projects that interest them and might benefit the company in some way. Google calls this practice "20 percent time." It is borrowed from university life, in which professors typically are encouraged to spend part of their time on projects of their own choosing.

"Work should be challenging," according to Google's corporate philosophy, "and the challenge should be fun." If the employees are satisfied, their work will be at a high level: "Give the proper tools to a group of people who like to make a difference and they will."

Despite their vast wealth, Page and Brin enjoy relatively simple lifestyles, though each man has a few favorite gadgets. Brin, for instance, likes to tool around on his Segway Human Transporter (an electric-powered scooter).

Both men shun publicity. When they do appear in public, however, they might be wearing bizarre

21st Century Content

Brin and Page decided that their new company would follow a completely fresh approach, evading corporate tradition. They have demonstrated that the traditional corporate model, where profit takes precedence over everything else, isn't the only route to success. Turning Google into a company where people really love their jobs has helped them attract the best possible employees.

Larry Page delivers a speech at the 2006 International Consumer Electronics Show in Las Vegas.

costumes and behaving flamboyantly. In 2006, Page gave the keynote speech at the International Consumer Electronics Show in Las Vegas. He came onstage riding on the back bumper of a robotic SUV, wearing sneakers, jeans, a T-shirt, and a white lab coat. At an event for their employees, Page and Brin wore capes and leaped onstage as if coming in for a landing.

Brin has published more than a dozen academic papers and has appeared on the *Charlie Rose Show* and other TV programs. Both men were interviewed on the Biography Channel in 2004.

Page has often said that he "has always wanted to change the world." Google has done that, to an extent far greater than most products or services. As author John Battelle put it, Brin and Page have "clarified and cleaned up the clutter of the Internet." While doing so, Google has "fundamentally changed the relationship between humanity and knowledge."

Brin has said he considers Google to be like a child who's just finished first grade, with a long and uncharted life ahead. If that's true, there's no telling what we can expect from Google as the 21st century rolls onward.

What's next for Google? We'll just have to wait and see!

Timeline

1973 Lawrence E. (Larry) Page is born in East Lansing, Michigan, in March. Sergey Mikhailovich Brin is born in Moscow in the Soviet Union (now in Russia), in August.

1978 The Page family gets a home computer.

1979 Sergey's family emigrates to the United States to escape persecution of Jews in the Soviet Union.

1982 Sergey receives his first computer as a birthday gift.

1989 While still in high school, Sergey enters the University of Maryland.

1991 Page graduates from East Lansing High School and enrolls at the University of Michigan, majoring in computer science and engineering.

1993 Brin earns his bachelor's degree from the University of Maryland and enrolls as a graduate student at Stanford University, pursuing a doctorate in computer science.

1995 Page and Brin meet for the first time during a tour for prospective Stanford University graduate students; Page earns a bachelor's degree in engineering.

1996 Page and Brin develop BackRub, a primitive search engine that uses the concept of "back links" to evaluate indexed Web sites.

1998 On September 7, Brin and Page found Google Inc.; Brin is president and Page is CEO.

1999 Googleplex headquarters is established in Mountain View, California.

2001 Eric Schmidt is hired as Google CEO; Brin and Page get new titles; Google Inc. earns a profit for the first time.

2004 In August, Google makes its initial public offering of stock.

2006 Larry Brilliant is hired to head new Google.org, created for charitable activities; Google acquires YouTube.

2007 In January, *Fortune* magazine names Google Inc. the Best Company to Work For in America.

Glossary

algorithm (AL-guh-rith-uhm) a mathematical formula for solving a problem using a specific series of steps

anti-Semitism (AN-tee-SE-mah-TI-zuhm) hostility toward Jews as a religious, ethnic, or racial group

browsers (BROU-zurz) a computer program that enables you to surf the Internet

copyright (KOP-ee-rite) the exclusive right to produce, publish, and sell a book, song, or other work

corporation (kor-puh-RAY-shuhn) a group of people who are allowed to run a company as a single person

database (DAY-tuh-bays) a large collection of data organized and stored in a computer

dissertations (dis-ur-TAY-shuhnz) detailed and extensive studies of a subject; one is submitted as part of the requirement to obtain a doctorate degree

eclipse (i-KLIPS) to do a great deal better than; to exceed

emigrated (EM-uh-gray-ted) left one's own country in order to live in another one

endowment (en-DOU-muhnt) a large amount of money meant to continue the operation of a company or foundation

entrepreneur (ON-truh-pruh-NUR) someone who operates and assumes the risk of a business

equity (EH-kwuh-tee) the value of an owner's interest in something; the common stock of a corporation

extrovert (EK-struh-vurt) someone who is outgoing and talkative and enjoys being with other people

hierarchy (HI-uh-rahr-kee) the ranking of people within an organization

philanthropy (fuh-LAN-thruh-pee) active effort to help others by giving time or money

prodigies (PRAW-duh-jeez) children who are unusually intelligent or talented

venture capitalists (VEN-chur KAP-uh-tuh-lists) people who invest in new or fresh businesses

For More Information

Books

Battelle, John. *The Search: How Google and Its Rivals Rewrote the Rules of Business and Transformed Our Culture.* New York: Portfolio, 2005.

Vise, David A., and Mark Malseed. *The Google Story.* New York: Delacorte Press, 2005.

White, Casey. *Sergey Brin and Larry Page: The Founders of Google.* New York: The Rosen Publishing Group, 2006.

Web Sites

Academy of Achievement: Sergey Brin & Larry Page
www.achievement.org/autodoc/page/pag0int-1
Includes a brief profile of the company, biographies of its founders, an interview with them, and images

Google Corporate Information
www.google.com/intl/en/corporate/index.html
For information about the company, its mission and philosophy, brief biographies of selected employees, and a timeline

How Stuff Works: How Internet Search Engines Work
computer.howstuffworks.com/search-engine.htm
Features a step-by-step explanation of how search engines work

INDEX

advertising, 22–25, 35

Battelle, John, 43
Bechtolsheim, Andy, 20–21
Bill and Melinda Gates Foundation, 31
births, 5, 7
Brilliant, Larry, 30–32
Brin, Eugenia, 7
Brin, Michael, 7

childhood, 4, 7–9

education, 5, 8, 9, 10–13, 19
employees, 22, 27, 38–41

Filo, David, 20

Gates, Bill, 31
Gates, Melinda, 31
Google Book Search, 33–34
Googlebot, 13
Google Foundation, 32
Google Grants, 35

Google.org, 30–32
Googleplex, 22, 38–41

initial public offering (IPO), 25, 26–27, 29
investors, 20–21, 22, 26, 37

Jobs, Steve, 21

media coverage, 22, 28, 36, 37, 38, 41–42

offices, 13, 19, 22, 38–41

Page, Carl Victor, 5–6, 11
Page, Gloria, 6
philanthropy, 29–35
profits, 23–24, 25, 30, 37, 41

Sandberg, Sheryl, 30
Schmidt, Eric, 26, 37
search engines, 13, 14–15, 19
shareholders, 27–28, 29–30, 37

Silicon Valley, 11, 17, 20, 21, 27, 30
Silverstein, Craig, 21
Sirotta, Milton, 19
Stanford University, 11, 15, 17, 27, 34, 37
stocks, 26–27, 29, 30, 32

Tesla, Nikola, 8–9

Venter, Craig, 34–35

wealth, 4, 29
Winograd, Terry, 37
World Wide Web, 13, 14–15, 19
Wozniak, Steve, 21

Yahoo!, 11, 18, 19, 20, 22
Yang, Jerry, 20
YouTube, 28

ABOUT THE AUTHOR

James M. Flammang is a journalist and the author of two dozen books. He writes about technology, business, and consumer issues. His specialties have ranged from cars to electronic gadgets to computers. Because of his interest in ideas that affect people's lives, and in the people who develop big ideas, Flammang enjoyed researching the lives of the two founders of Google. This is his third book for young readers. Flammang lives in Chicago, but travels extensively for both business and pleasure. His own Web site, *Tirekicking Today*, has been at www.tirekick.com since 1995.